CAREER AS A

CARPENTER

CARPENTRY CONTRACTOR

WHAT DO BERNIE SANDERS, HARRISON FORD, and Matt LeBlanc have in common? They all worked as carpenters before becoming famous in movies and politics. Most carpenters are not hoping for fame, nor are they working in their craft temporarily while planning for another career. Carpenters love what they do and are happy to build rewarding careers that will last a lifetime.

Carpentry is a craft that primarily involves making things from wood. Most carpenters use their skills to build and repair residential and commercial buildings. Within the construction industry, there are several types of carpentry, each requiring different specialized skills. The two main types are rough carpentry and finish carpentry. Roofers and framers are rough carpenters. Their work is rarely seen. Finish carpenters work on all the fine details that will be seen, like trim, molding, and fixtures.

Not all carpenters work in the construction industry. Some build bridges and ships, while others make furniture or boats. Some even create theatrical sets for movies and TV.

Within the construction industry, carpentry projects can vary widely from one project to the next. However, most involve the same basic steps. It starts with reading blueprints and other instructions provided by supervisors or homeowners. From there, carpenters carefully measure, mark, and organize materials. The materials are cut and shaped with hand and power tools, then joined together with nails, screws, staples, and glue. Levels, plumb bobs, and framing squares are used at every step to make sure everything is straight and smooth.

Sometimes carpenters use prefabricated components rather than creating pieces from scratch. Installing factory-made staircases, wall panels, pre-hung windows, and roofing assemblies is quicker and easier than cutting and assembling many small pieces.

Carpenters work in every city and community because they are needed everywhere. Once they are fully trained, they can find employment anywhere in the US. Those who live in cities often work for large construction companies that hire crews of dozens or even hundreds of carpenters, each of whom is assigned to a specific task. Carpenters working in smaller communities are less likely to specialize and will usually put a broader range of skills to good use. They are typically employed by small contractors and residential builders. There are also many carpenters who are self-employed. In fact, one out of three carpenters is an independent contractor who usually seeks work directly from homeowners.

There is a high demand for carpenters. The number of positions available for new carpenters is on the rise. Because of the high turnover and the never-ending need for more buildings, there will always be jobs for those who want them. The prospects are excellent for those entering the field, however, job opportunities are best for well-trained carpenters with diverse skills. These skills are often learned on the job while working as apprentices or helpers to more experienced carpenters. No college is required, but it still can take three to five years to complete an apprenticeship.

A career in carpentry has many attractive features, including excellent pay, easy entry, fulfilling work, flexibility, self-employment options, upward mobility, and good job outlook. If you are looking for a career with minimal stress and good work-life balance, that does not require a college degree, read on. Carpentry may be what you are looking for.

WHAT YOU CAN DO NOW

YOU DO NOT NEED A COLLEGE DEGREE to become a carpenter, but you need to graduate from high school or earn an equivalent. Make good use of your time in high school by selecting courses that will help prepare you for a career in carpentry. First, make sure to fulfill any requirements for graduation. Then, consider the following classes:

- Wood shop
- Mechanical drawing and drafting
- Algebra and geometry
- Blueprint reading
- Sketching
- General vocational training
- Physical education

Consider learning Spanish. Employers value carpenters who can communicate with their Spanish-speaking workers. That skill alone can help you advance in your career more quickly.

Many carpenters are self-employed contractors. That means owning and operating a small business. Useful subjects to study would include computer applications, marketing, business management, and any classes that would improve communications skills.

Although most carpenters learn their trade through an apprenticeship, some learn on the job, starting as a helper. You are not eligible to apply for a formal apprenticeship until you are 18, but if you are at least 16 years old, there are plenty of opportunities for helpers and beginners at nonunion construction companies and contractors, especially during the busy summer months. A part-time job after school or during the summer will provide the kind of hands-on experience that will be essential for getting into an apprenticeship program in the future. It can also lead to full-time work when you graduate.

A good way to explore whether this is the right career for you is to shadow a carpenter for a day (or more) on the job. Just about any carpenter should be willing to do this for you – especially if you offer to help out by carrying tools and materials or cleaning up. Free help is hard to turn down! Be sure to ask plenty of questions and gather advice on things you can do to prepare.

Find out if your school's drama department offers a set building class or opportunities to build the scenery for school plays. It is an interesting way to learn the basics of carpentry while also allowing you to be a bit more creative than most shop classes.

Find out if there are any carpentry classes for beginners in your community. Look for free or inexpensive classes sponsored by your local community center or city vocational college. You can also use Google to help you find any carpenters in your area who may be putting on workshops or any classes you could attend.

Work experience is valuable regardless of whether there is actual pay involved. Volunteer work will help you gain experience and it looks good on a résumé. Groups like Habitat for Humanity sponsor construction projects all the time and are always open to volunteers with or without experience in the construction of buildings.

HISTORY OF THE CAREER

CARPENTRY IS ONE OF THE FIRST REAL PROFESSIONS, dating back thousands of years. Wood has always been a readily available building material that has been used for one of mankind's primary needs – shelter. Archaeologists have found evidence of wood construction that is over 7,000 years old. Excavations in eastern Germany turned up water well casing built of split oak timbers from the early Neolithic period that utilized joining techniques such as mortise and tenon and notched.

Ancient Egyptians used similar joining techniques during the time of the first pharaoh, about 3100 BC. Egyptian drawings as old as 2000 BC illustrated how they also used pegs, dowels, and leather or cord lashings to strengthen these joints while making various kinds of furniture. A variety of tools was used, including adzes, chisels, bow drills, pull saws, and axes. Ancient Egyptian carpenters regularly practiced their craft and developed techniques that continually advanced the trade. For example, animal glue was introduced during the 16th century BC. They later invented the art of veneering, gluing thin slices of wood together – an art that is common to this day.

The evolution of carpentry is directly related to improvements in carpentry tools. One of the best-known contributors to the development of carpentry tools was Lu Ban (507-440 BC), a Chinese carpenter, engineer, and inventor. Lu Ban is known for being one of the originators of woodworking in China and is often referred to as the patron saint of Chinese builders and contractors. He is credited with introducing the plane, chalk line, and other tools that are still in common use today. During the Ming dynasty, about 1500 years after Lu Ban's death, his instructions regarding architecture were compiled in the book, *Lu Ban Jing*.

Unfortunately, wood is not the most enduring building material. Buildings are often lost to fire, floods, wood rot, war, and general deterioration. However, there are a few remarkably old examples still standing. One is the Great Buddha Hall of the Nanchan Temple in China. It was built around the year 782 BC during the Tang Dynasty, and is China's oldest preserved timber building. Another example is the Greensted Church in England, built in the 11th century. Only parts of that building are still standing. In Europe, the oldest examples are stave churches. These medieval wooden church buildings were once common in northwestern Europe. The only surviving stave churches are located in Norway where they were built in the 12th and 13th centuries.

Carpentry is a trade that is unique in terms of the way it is taught. Many professions have been taught in classrooms or through textbooks for centuries, but the knowledge and skills of carpentry have been historically passed down from person to person. There is little instruction about carpentry prior to the invention of the printing press in the 15th century. One of the only exceptions is the *De Architectura* (Ten Books on Architecture) from the first century BC. This guide for building projects, which was written by Roman architect Marcus Vituvius Pollio, discusses how to plan and erect many different kinds of buildings. It was not until the 18th and 19th centuries that carpenters began to see regularly published guides and pattern books.

The Middle Ages

The Middle Ages (also known as the Medieval period) spanned a thousand years that started with the fall of Rome and ended with the Renaissance in the 15th century. Throughout this time, wood was the most common building material. Carpenters were held in high esteem, widely considered to be among the most skilled craftsmen. Growing communities were dependent on carpenters – in fact, the availability of good carpenters often meant the difference between the success and failure of new communities. Coming up with enough new carpenters was not easy. Training was provided through apprenticeships, and apprenticeships were only offered by established carpenters willing to pass on their knowledge to a young man (often a son). It was typically a long process, informal and unstructured, and it could take years.

Eventually, a system of formal apprenticeship was developed, supervised by craft guilds and town governments. Carpenters, like many other craftsmen, were then required to belong to the guilds. Eventually, a carpenters union was formed, but that was not until 1881.

Industrial Revolution

Until the 16th century, buildings were made from timbers. Once logs could be cut into lumber that changed. Sawmills sprang up all over Europe and housing carpentry started to change dramatically. Carpentry advanced again during the Industrial Revolution with the production of cut nails. No longer held back by the slow, tedious process of making handmade nails, more modern methods of carpentry led to what is practiced today. Further refinement came in the 19th century, with the development of electrical engineering, hand-held power tools, chemically treated lumber, wire nails, and mass produced screws.

Modern Carpentry

Today, carpentry is a multifaceted trade. Many carpenters still frame houses, but there are numerous specialties within the career. Carpenters can now focus solely on making cabinetry, furniture, roofs, decks, shutters, or cupolas. They can do anything from rough framing of a commercial building, to fine finishing, to restoring historic buildings. New technologies have helped carpenters work with more precision and speed than ever, but the basic techniques of the trade have been passed down for centuries. Many carpenters are considered artisans, transcending functional skill levels to create wooden buildings and furnishings of unique beauty.

WHERE YOU WILL WORK

THERE ARE MORE THAN A MILLION CARPENTERS at work in the US. Most work in the construction industry, where they account for the largest share of the building trades. About 20 percent of all carpenters work in residential construction, building houses. Another 12 percent work in nonresidential building construction. This includes building warehouses, recreational facilities, bridges, churches, shopping centers, restaurants, stadiums, and any other commercial structure. The third biggest source of employment for carpenters is building finishing (working on nonstructural parts of a building). There are also many carpenters who are hired to do repairs or renovations. Not all carpenters work in construction. A carpenter can work with wood to make anything from furniture to boats to musical instruments.

The majority of carpenters – over 60 percent – are employed through unions. In the construction industry, the second largest group of employers is contractors, both general contractors and carpentry contractors. Outside of construction, manufacturers, government agencies, retail establishments, and a variety of other industries employ carpenters.

Carpenters are at work everywhere. Jobs are generally distributed in proportion to the population. Where there are many buildings, many carpenters are at work. That means there are more carpentry jobs in cities than in sparsely populated areas.

Most carpenters work locally, moving from one project to the next within driving distance of their home. There usually is not much opportunity for travel, though some large companies do send workers to other locations on temporary assignments.

Self-employment is common in carpentry. About one of every three carpenters is self-employed. Most of these self-employed workers do some sort of residential construction and are often employed directly by homeowners. It is not easy keeping a steady flow of work coming in. Self-employed carpenters often have to alternate between working for a contractor and working as contractors themselves on small jobs, depending on the availability of work.

Work Environment

Working conditions for carpenters vary. Some carpenters work in cramped indoor spaces, while others are positioned high off the ground on roofs or other structures. Because carpenters do so many different types of work, from installing kitchen cabinets to building bridge supports, they work both indoors and outdoors. Those who work indoors are usually able to enjoy a steady, normal 40-hour week. The hours for those working mostly outdoors will depend on the demands of the project. Weather is another big factor. Extreme temperatures or stormy weather can cause delays and limit a carpenter's ability to work.

Most carpenters work full time, but like most careers in construction, scheduling can ebb and flow with the seasons. There are peak periods, usually during the warmer months, that may require working evenings and weekends. The amount of overtime depends on the type of construction and geographic region, and will vary from one project to the next.

THE WORK YOU WILL DO

CARPENTERS USE WOOD AND OTHER MATERIALS to construct and repair building frameworks and structures. When working on buildings, whether residential or commercial, this includes stairways, doorframes, window frames, partitions, and rafters. In addition to buildings, they also construct supports for large projects such as bridges, roads, piers, and water vessels. Carpenters also work with many pre-constructed components. They install things like kitchen cabinets, siding, and drywall, as well as prefabricated windows, doors, bathrooms, and wine cellars.

Carpenters who have gone through apprenticeships are trained to do the following:

- Read and follow blueprints and building plans.
- Make sure local building regulations will accommodate client needs and wants.
- Properly use a variety of hand and power tools while following safety procedures.

- Measure, cut, and shape wood, plastic, and other materials.

- Create frameworks for doors, windows, walls, and floors.

- Install pre-made structures and fixtures, such as windows, doors, and molding.

- Erect, level, and install large building framework with the aid of rigging hardware and cranes.

- Inspect, repair, and replace damaged framework and fixtures.

The work of carpenters is diverse and every job is different. However, most carpenters work in the construction industry, building residences, factories, power plants, and all types of commercial and public buildings. Construction projects follow the same basic procedures. The first step is to read the blueprints and get a clear understanding of the immediate task and how it fits into the overall layout. Next, the carpenter measures, marks, and organizes the materials that will be used. The materials are then cut and formed. The carpenter will connect all of the pieces together using adhesives or nails, depending on the material. The last step is to review the work to make sure everything is straight and smooth, and fits like it is supposed to.

Some jobs do not require all of these steps, particularly when pre-made pieces are used. For example, stairs are often prefabricated so there is no cutting and assembling needed. Carpenters would only have to install the staircase, which can usually be accomplished in a few simple steps. However, pre-made components do not always fit properly. In that case, the carpenter would have to make adjustments to the framework to make it work.

Carpenters use many different hand and power tools. Most employers expect carpenters to have their own tools. In fact, the salary often depends on the number of tools a carpenter brings to the job. The most important tool is a simple one: a tape measure. It is used on nearly every project to make sure that the pieces being cut are the proper size in order to avoid waste of time and money. Rulers, squares, line chalk, and levelers are used to ensure pieces are straight and level.

A variety of tools are used for cutting and shaping wood, plastic, fiberglass, and drywall. They include hand and power saws, chisels, drills, and sanders. To fasten materials together, carpenters use hammers, nail guns, power staplers, screws, and adhesives.

A Versatile Career

Carpentry is one of the most versatile of all construction occupations. For example, some carpenters specialize in installing kitchen cabinets or drywall in homes, while others insulate office buildings, or construct concrete forms for footings or pillars. There are many possibilities and carpenters typically choose one area to work in for most of their career. Here are a few examples:

New Home Construction

Residential carpenters specialize in building single-family homes, townhomes, and condominiums. Some work on every aspect of the project, from start to finish. They might start by building forms for concrete footings, walls, and slabs, then framing walls, floors, and roofs. They would finish by installing stairs, drywall, doors, windows, molding, and cabinets. Other carpenters prefer to focus on one type of work, such as laying wood floors or building decks.

Remodeling

Fully trained carpenters can switch from building new homes to remodeling and renovation of existing structures. This requires a variety of skills that go beyond using a hammer and nails. This is especially true in older buildings where components and fixtures may not be replaceable. There are many historic restoration projects today, particularly in cities that are gentrifying entire neighborhoods of old buildings. Carpenters who can handle this type of work are highly valued, and those who have well-rounded skills have an edge when it comes to getting contracts.

Commercial Carpentry

This type of carpentry involves building and retrofitting office buildings, hospitals, hotels, schools, and shopping malls. Commercial carpenters perform many of the same tasks as residential carpenters, however, because these are large projects, most specialize in one particular type of work. This could be framing interior partitions, building curtain walls (a wall that encloses a space within a building, but is not a supporting wall), installing wooden concrete forms for footings, or erecting shoring and scaffolding for tall buildings.

Industrial Carpentry

Carpenters in this area typically work on civil engineering projects or in industrial settings. Basic tasks include building scaffolding and creating and setting forms for pouring concrete. This is not like building a form for a patio. Forms in industrial carpentry are used to build tunnels, bridges, dams, power plants, and sewers. Some industrial carpenters build tunnel bracing or partitions in underground passageways and mines to control the circulation of air to work sites.

Independent Contractors

Large construction companies or small general contractors employ most carpenters. However, many carpenters prefer to work for themselves. About one out of three choose to be independent and be their own boss. Of all the trades within the construction industry, that is the highest percentage of self-employed people. There is a high demand for skilled carpenters, so there are excellent opportunities for those going into business for themselves.

Self-employed carpenters range from one-man independent contractor operators to general contractors with dozens of carpenters on the payroll. Most are somewhere in between, small contractors with one to three carpenters and helpers. Most are involved in residential construction, both new housing and remodeling.

Not all independent carpenters work in construction. Some specialize in installing a variety of materials like glass, tiles, or carpets. Others create, repair, and restore furniture, cabinets, and hardwood floors.

Self-employed carpenters have the potential to earn more money than their employed counterparts. However, they have more to do than just carpentry work. They have to continually market their services, attract new customers, write and submit bids to potential customers, order materials and keep track of inventory, and instruct and manage employees and subcontractors. There is considerable paperwork involved, from doing quarterly tax returns to getting permits from local building departments. It can become too much for some carpenters. They might decide to compromise by taking on smaller side jobs on weekends while working for someone else during the week.

CARPENTERS TELL THEIR OWN STORIES

I Am a Freelance Carpenter

"I have been making stuff out of wood since I was a little kid. Today, I'm a working journeyman carpenter and licensed builder. I work alone, taking on mostly remodeling projects, but also custom cabinetry, finish trim work, custom doors and stairways, and historic renovations. I struck out on my own about five years after completing an apprenticeship. I made the switch from employee to independent contractor because I love doing the kind of high-end work that a certain segment of homeowners appreciate and are willing to pay for.

As a business owner, I am continually seeking new clients and new projects. This very often means I am on my smart phone as much as my miter saw. I get job leads through social media, but the best source of new clients is my network of past clients. Referrals and word-of-mouth advertising are far more effective than any kind of paid advertising.

The best part of being a freelancer is the freedom to work when, where, and how much I want. But of course, there is a price to pay for that independence. It is up to me, and me alone, to keep the work flowing in. Plus, I have to manage my time since I do all the work myself.

My idea of carpentry is a combination of art and science. I think it is a good career choice for people who view this work as a serious profession rather than something one might do temporarily to make some money."

I Am a Union Carpenter

"I never had a desire to go to college. Because of my experience working as a carpenter's helper during the summer, I decided the best choice for me was to join the union. That was more than 20 years ago. Since then, I have worked in both labor and management positions. I have never regretted my choice.

Anyone with the desire to become a carpenter can do so. However, there are advantages to being involved with the union. For starters, the training is extremely thorough, but it does take a few years. It is designed for people who want a career, not just a job. The wages and benefits are excellent – better than you can expect from most nonunion employers. There is also the benefit of steady work. Union representatives work with local and state workforce development boards as well as large companies and government agencies to get the biggest and best contracts. Little independent guys can't get those contracts. They simply don't have the means to fulfill them.

One way to get started as a union carpenter is to apply to a local union. New applicants are put on a list and when there is an opening in the apprenticeship program, they are tested. Another way is to apply directly to union contractors. If new carpenters are needed, successful applicants will be entered into the program. Upon completing an apprenticeship, a carpenter becomes a journeyman. At that point, he can travel and work anywhere in the US and Canada."

PERSONAL QUALIFICATIONS

DO YOU ENJOY WORKING WITH WOOD? ARE YOU CREATIVE? DO YOU LIKE BUILDING THINGS? If the answer to these questions is yes, then a career in carpentry could be what you are looking for. To be a successful carpenter, you primarily need to be good with your hands and have the ability to see the big picture and not just the immediate task. Beyond that, there are other key characteristics shared by successful carpenters:

Detail Oriented

Carpenters perform many tasks that are important in the overall building process. Making precise measurements, for example, may reduce gaps between windows and frames, limiting any leaks around the window. An incorrectly measured joist could affect the safety of an entire building. Carpenters also need the ability to look at a wooden structure and assess any problems with it.

Physical Fitness

Carpentry work can be strenuous. Strength is needed to carry tools and heavy materials – one plywood sheet alone can weigh 50 to 100 pounds! Endurance is needed to stand, climb, and bend for long periods. Carpenters are on their feet most of the day doing active physical labor. A good sense of balance is also necessary.

Dexterity

In addition to strength, great hand-eye coordination is needed. Carpenters use many tools, some of which can cause serious injury if misused. Expensive materials can be damaged. Simply hitting a nail on the head with accuracy, for example, is crucial to avoid getting hurt or damaging wood.

Knack for Problem Solving

Problem-solving skills are especially important for projects involving repairs or remodeling, but even new projects can pose numerous problems. A special order prefabricated staircase arrives at the work site and it does not quite fit. Boards break, tools stall, materials show up late or they are the wrong kind or quantity. An inspector is having a bad day and is determined to find every minor flaw in the work. There will always be a glitch somewhere, in spite of the best-laid plans. A good carpenter will figure it out and come up with a quick, effective solution. Successful carpenters enjoy the challenges that come with the territory. At the end of the day they feel good about what they have accomplished.

Interpersonal Skills

Carpenters usually work alongside other craft workers. Those who are self-employed deal directly with customers and suppliers. Good communications and strong interpersonal skills are needed. It pays to keep a pleasant demeanor and make every effort to get on well with other workers and customers.

A Head for Business

Self-employed carpenters need more than basic carpentry skills. They must be able to attract new customers, track inventory, manage workers, maintain accounts, and plan projects from beginning to end. Math skills, in particular, are needed by all carpenters on a daily basis to calculate measurements of materials to be cut. Carpentry contractors also use math to estimate materials and labor costs, bid on new jobs, and ensure that projects stay on budget.

ATTRACTIVE FEATURES

A CAREER IN CARPENTRY IS LIKE ANY other profession – it has its pros and cons. Overall, carpenters enjoy a job with minimal stress, good work-life balance, and solid prospects for advancements with increases in earnings. Most carpenters say they are very satisfied with their career choice because the pay is good, there are ample employment opportunities, and they like the flexibility. Here is a closer look at some of the many attractive features a career in carpentry has to offer:

Excellent Pay

A carpenter's earnings are good by any standard, but for a career that does not require a college degree, they can be outstanding. The annual median wage is about $45,000, and experienced carpenters can earn more than $75,000. How much any carpenter earns is, of course, dependent on the level of the work done and the experience of the individual. The opportunity exists to improve skills and receive pay raises.

Fulfilling Work

It is always desirable to make good money, but when it comes right down to it, who wants to go to a job they dislike every day? More than 80 percent of carpenters give their career a high job satisfaction rating. Many even view it as their calling. There is something special about building something that you know is going to be around for a long time. The finished work of a carpenter is very tangible. It can be seen, touched, and enjoyed by all kinds of people – not just the carpenter. In spite of all the hard labor that goes into a project, for someone who takes pride in the work, this knowledge can be quite gratifying.

Good Job Outlook

Carpenters have experienced job growth year after year, and that is not expected to change over the coming decade. The main reason for increased job opportunities is the number of new homes being built and the popularity of home remodeling. Carpenters can also expect to find jobs when they want them because there is a wide range of work throughout the year. Home building, which generally takes place in the warm, dry months, is only one type of project a carpenter can do. Other jobs include bridge building, cabinet making, furniture making, TV and movie set building, and more. Carpenters do not just work on construction sites. They also work in factories, lumberyards, shipyards, and anywhere else wood is being crafted.

Flexibility

Unlike many workers, carpenters are not stuck doing the same type of work or operating the same machine every day. A carpenter's work is different from one project to the next, and the work varies from one day to the next. Boredom is never a problem. Even in basic carpentry, there is no need to worry about tedious and repetitious job tasks. Carpentry work starts and stops with each project. If a project is unsatisfying, you can always move on to the next. Carpenters are not tied to a desk or factory work station either. They enjoy very mobile and active lives, taking advantage of slow times to go on vacations or taking up small side projects that align with their personal interests.

Self-Employment Options

Carpenters tend to be an independent bunch and many reject the idea of reporting to a boss every day. For them, being self-employed is the way to go. Going into business is typically expensive and difficult, but this is not so in carpentry. Once you have acquired the necessary skills and tools, you can set up your own carpentry business. One out of three carpenters do this at some point in their career. It is a great option for someone who is self-motivated and wants the freedom to pick and choose projects. It is especially important, however, to provide good work for clients. Self-employed carpenters depend on recommendations and referrals from customers to keep future projects coming their way.

UNATTRACTIVE ASPECTS

DESPITE THE STRONG UPSIDES, there are also disadvantages when choosing a career in carpentry. First, there is the physical aspect. It is, after all, a physical occupation. Carpenters spend most of their time on their feet, standing, climbing, bending, and kneeling. There are also heavy materials, tools, and equipment to carry. All in all, it can be physically exhausting. Then there is the weather to deal with. Most carpenters are outdoors people, but working outside when it is hot, cold, drizzly, or worse, is no fun. If you dislike the outdoors and do not like being exposed to uncomfortable conditions, you might want to consider a different career path.

Carpentry is potentially dangerous. In fact, it has one of the highest rates of injuries and illnesses of all craft occupations. The most common injuries include muscle strains from lifting heavy materials, bruises and broken bones from falling off ladders or slipping on wet surfaces, and cuts from sharp objects and rough materials. A carpenter's tools can also pose a risk if not handled properly. Electric saws, drills, and other tools can cause serious injury if misused or mishandled. There is also the possibility of being exposed to harmful materials like asbestos. Such a possibility is becoming less likely as this material has not been used in construction for many years. Only those who are rehabbing older homes need to be on the lookout for asbestos. All these occupational hazards can be avoided by taking proper safety precautions and following work regulations. Safety procedures are addressed repeatedly throughout apprenticeship programs.

In addition to being physically demanding, carpentry is also mentally challenging. For example, carpenters must be able to visualize the completed project and see how the task at hand fits into the big picture. Carpenters also use their math skills every day and often have to make quick calculations on the fly. There is a saying in carpentry: measure twice, cut once. By sticking to that rule, most carpenters avoid making mistakes.

Like most construction jobs, carpentry is subject to economic conditions. When the economy takes a downturn, housing projects are often put on hold. During these times, carpenters can experience slow times. Conversely, when the economy picks up, so does construction and everything connected to it, such as furniture and kitchen remodeling. At such times, you may be swamped with work and not be able to accommodate all your potential clients. Regardless, there will always be smaller projects that carpenters can pick up to stay busy year around.

Apprenticeship programs are lengthy. It can take four years to complete an apprenticeship. That is about the same time it takes a full-time student to earn a college degree. There is a big difference, however. College students pay to go to college, while apprentices are paid while they learn. The pay is not much – but it is about half what they will earn when they complete their training and become journeymen. It is more than minimum wage and enough to pay the bills while learning a valuable trade.

EDUCATION AND TRAINING

THERE ARE SEVERAL WAYS FOR CARPENTERS to learn the craft. One way is to get hired as a carpenter's helper or find an employer who will provide informal on-the-job training. This is a good way to go for someone who is eager to get started and does not want to spend time in a classroom. The disadvantage is the narrow scope of training that might be provided. Overall, how broad and comprehensive the training will be depends mostly on the size of the employer. For example, a small, independent contractor will probably be working exclusively on residential construction. The new carpenter's helper will likely be taught how to do rough framing, but not much else. On the other hand, a large contractor might be able to offer training in a variety of areas, such as finishing, cabinet installation, or drywall installation. Either way, an experienced carpenter will teach the beginner one skill at a time as needed to complete projects. In the long run, it is beneficial to gain as many skills as possible.

Another way to get started is to take classes in order to learn the basic skills of the trade before looking for employment. There are many two-year technical schools and community colleges that offer carpentry certificates or even degrees, and many of those are affiliated with unions or contractor organizations. There are also online certificate programs. For example, The National Association of Home Builders offers Pre-Apprenticeship Certificate Training (PACT) through the Home Builders Institute.

After obtaining a degree or certificate, there are two choices: find an employer who will offer informal on-the-job training or become an apprentice. Most professional carpenters agree that the best way to learn the trade is through apprenticeships because they usually provide the most comprehensive training. That is not always possible though, because there are a limited number of apprenticeships available. That is the main reason employer-provided informal training is more common.

Apprenticeships can be informal or formal, union or nonunion. They can be found in most cities and are sponsored by local union branches and contractor associations. Local unions like United Brotherhood of Carpenters and Joiners of America, and National Association of Home Builders sponsor apprenticeships or run similar programs. Other programs that provide instruction and practical experience are run by local branches of Associated General Contractors and Associated

Builders and Contractors.

The basic qualifications for a person to enter an apprenticeship program are as follows:

- Minimum age of 18
- High school education or equivalent
- Physically able to do the work
- US citizen or proof of legal residency
- Pass substance abuse screening

Most apprenticeship programs take three to four years to complete. During those years, apprentices must complete at least 144 hours of technical training in the classroom plus 2,000 to 8,000 hours of paid on-the-job training. In the classroom, apprentices learn carpentry basics, including blueprint reading, building code requirements, mechanical drawing, math, and safety and first aid practices. They are also taught basic design principles and the basic procedures of carpentry work like framing, interior and exterior finishing, and layout. All carpenters must pass Occupational Safety and Health Administration (OSHA) safety courses.

Outside of the classroom, apprentices gain practical experience in real work settings. They learn how to properly use carpentry hand and power tools, materials, and equipment. They also learn how carpentry fits in with other construction jobs. Some apprenticeship programs also offer specialized training in creating and setting concrete forms, rigging, welding, scaffold building, working within confined work spaces, and fall protection.

An apprentice becomes a journeyman after satisfactorily completing the apprenticeship program. There are usually tests, both practical and written, to determine if journeyman status should be awarded.

Advancement

Advancement opportunities for carpenters are better than for other construction professions because they are involved in all phases of construction. They might become first-line supervisors, general construction supervisors, or inspectors. Carpenters who want to advance their careers often take additional training provided by associations, unions, or employers.

Some experienced carpenters advance by going into business for themselves. They may start out as independent contractors, working alone. As their business grows, they employ others to do the actual carpentry work while they focus on managing and building the business.

EARNINGS

CARPENTERS ARE WELL PAID. IN FACT, with a median annual income of almost $45,000, they are among the highest paid of the many construction occupations. Carpenters are usually paid on an hourly basis rather than a yearly salary.

Only the lowest 10 percent of carpenters earn less than $27,000 a year. Most of those are beginners in nonunion jobs. The starting pay for apprentices is generally between 30 percent and 50 percent of what fully trained carpenters make. Wages increase as apprentices learn to do more and their skills improve. At the other end of the scale, the highest 10 percent earn more than $75,000.

Wages for carpenters can vary widely depending upon geographic location, the industry, the size of the employing company, and whether the company is union or nonunion. Wages are higher in metropolitan areas than rural areas or small towns. The highest paid carpenters are located in cities in California, Hawaii, and Alaska.

Pay rates also vary among different industries and types of work. For example, the motion picture and video industry is known for offering the highest pay for carpentry work, but that is a relatively small industry that can be hard to break into. Here are the median annual wages for carpenters in the industries that employ the largest numbers of carpenters:

- Nonresidential building construction
  $47,000

- Building finishing contractors
  $44,000

- Residential building construction
  $41,000

Union carpenters are usually paid more than their nonunion counterparts. The only possible exception would be those who are self-employed or run their own contracting companies. Many carpenters belong to the United Brotherhood of Carpenters and Joiners of America, Associated Builders and Contractors, or Associated General Contractors of America. These big national unions have the power to negotiate the best deals for their members. However, wages will still vary among union locals. Although it can vary with the union local, most hourly wage rates are based on a regular 35 to 40 hour week. Actual total income for union members is a combination of basic wage rates, plus employer contributions to retirement funds and other employee benefits.

About a third of all carpenters are self-employed. Self-employed carpenters often work in residential construction. Although those jobs are not the highest paying, self-employed contractors often come out ahead because they are not sharing part of the profits with an employer.

Nearly all carpenters work full time. Overtime is common, especially when trying to meet project deadlines. Most carpenters do not mind working into the evenings, or on weekends or holidays because overtime pay is usually double the regular basic rate. Overtime may also be due to weather delays that limit the number of hours that can be worked during normal workdays. This happens most often during the height of the building season, when summer temperatures can get too hot to work outside or thunder storms cause work to be called off. Self-employed carpenters enjoy more flexibility than their employed counterparts, but the results of setting their own schedule and taking time off are loss of income.

Most full-time employed carpenters earn benefits. These typically include vacation, sick leave, and health insurance. Apprentice carpenters receive the same benefits as experienced journeymen carpenters, but only after an initial waiting period. Self-employed carpenters must provide their own insurance and retirement plans.

OPPORTUNITIES

THE NUMBER OF CARPENTRY JOBS will grow by about six percent over the next 10 years. That growth rate is about average when compared to all occupations. The number of job openings will vary by geographic area. Construction activity typically mirrors the movement of people and businesses. Areas that experience the largest population growth will require the most carpenters.

Most openings in the coming years will be created by carpenters who retire or change careers. Career changes are common in the carpentry field. Very often, people become carpenters because they would rather not invest in a college education. At some point, they realize the tradeoff is work that is physically demanding and they look for more comfortable employment. This results in more openings for new workers looking to become carpenters.

The housing market has been expanding year after year, and there is no slowdown expected for the foreseeable future. The Census Bureau reported 1,182 million building permits were issued in 2015. That is the highest number since 2008, which was just before the housing bubble burst. New home construction is the largest sector employing carpenters and many more will be needed for these projects. The surge is partly due to the rebounding housing market. In areas of high employment and high incomes, home buyers want larger and better homes, as well as vacation homes. Carpenters are also working on a large number of starter homes, apartments, and rental housing. Most of these residences are needed by the young adult population, who cannot yet afford large, expensive homes.

Not all carpenters work in new housing construction. Many homeowners prefer to remodel their existing homes rather than buy a new one. Real estate investors and landlords hire carpenters to remodel, restore, and repair homes. Many commercial buildings, restaurants, factories, and loft buildings are slated for restoration and modernization.

Beyond construction, there is the need to repair and replace roads and bridges. However, despite the clear need to upgrade existing infrastructure, opportunities for carpenters may be slower because of federal and state government budget constraints.

Although the industry is growing and there will always be a need for skilled carpenters, job growth is being somewhat restrained by automation. The use of prefabricated and manufactured materials, such as pre-cut stairs, doors, windows, and even complete bathrooms is on the rise. Contractors know it is easier and cheaper to use these materials rather than hire a carpenter to produce them from scratch. Carpenters are still needed to install these pieces, but much of the labor-intensive and time-consuming activities have been removed from the process. Complex components are also being produced in factories. For example, roofs that once had to be built with hammers, saws, boards, and nails, can now be installed in a single operation. There are other technological advances, such as modern adhesives and pneumatic tools, that affect the amount of time carpenters spend on tasks.

The construction industry, which is the primary source of employment for carpenters, is notoriously unstable. Projects are always temporary and the industry reacts quickly to economic cycles. The number of new buildings being constructed fluctuates with changes in interest rates, mortgage rates, weather, the time of year, and investment trends. It is not uncommon for carpenters to find themselves between jobs for periods of time. Experienced carpenters know how to maintain a stable income by gaining diverse skills and taking on a variety of small projects during the slow times.

GETTING STARTED

MOST CARPENTERS START OUT AS APPRENTICES. Apprenticeships are sponsored by labor unions, professional organizations, and employers. A union apprenticeship program is a formal training program that leads to union membership and good jobs. The first step is finding a local union for carpenters. You can locate your local union office through the United Brotherhood of Carpenters and Joiners of America website or by asking any carpenter. You will be asked to fill out an application and be interviewed like any other job. There is competition to get into union apprenticeships. If you have a friend or relative in the union, be sure and use them as a reference. There was a time when having a relative in the union was a requirement for entry. That is no longer the case, but old traditions still linger and you can use them to your advantage. Unions also give preferential consideration to people who have served in the armed forces, especially if they were trained in any of the building skills.

There are also carpentry apprenticeship programs offered by professional organizations such as Associated General Contractors, Associated Builders and Contractors, and National Association of Home Builders.

Your future does not have to depend on a union or professional organization. There may be carpenters in your area who will take you on as an apprentice. Look for carpenters, contractors, construction companies, and shipbuilding companies, and furniture companies, who may be looking for an apprentice.

There are also informal training situations to lead to full-time work. If you are unable to participate in an apprenticeship, look for opportunities to become an on-the-job helper. Unlike apprentices starting out, helpers are not necessarily expected to have much experience or knowledge about carpentry. Mostly, you need to be strong, willing, and able to do the work you are assigned. If you are enthusiastic and eager to learn, there is the potential for the person or company you help to be fully invested in teaching you the ins and outs of the carpentry trade.

There are numerous ways to find potential employers. Many employers send job opening announcements to vocational schools. If you are going to a trade school, be sure and check the job boards or talk to your counselor. You can look in the help wanted ads in your local newspaper, but you are better off going online. Check out job boards, specialty websites that cater to carpentry or general construction trades, and sites like Craigslist. You are basically looking for any business that involves carpentry of any kind. Google "carpentry contractors near me" and give each of them a call.   If they are not hiring any trainees, ask if they know of anyone who is.

Many carpenters who are just starting out find their first jobs through word-of-mouth. Carpentry is like any other business – it is a small world and contractors usually know what other contractors are doing. This can be a very effective way of getting to the right company quickly. Be sure and use the referrer's name. It does not matter that they do not know you, and did not actually recommend you. Just having a name to use breaks the ice and gets their attention. When you ask for an interview, point out what you have to offer – even if it is only a good attitude and desire to learn the trade. If you have any experience or have taken any carpentry or building courses, be sure to mention that.

ASSOCIATIONS

■ **Associated Builders and Contractors**
http://www.abc.org

■ **Associated General Contractors of America**
https://www.agc.org

■ **Home Builders Institute**
http://www.hbi.org

■ **United Brotherhood of Carpenters and Joiners of America**
www.carpenters.org/Home.aspx

■ **National Association of Home Builders**
https://www.nahb.org

PERIODICAL

■ **Builder**
http://www.builderonline.com/tag/carpentry

WEBSITES

■ **The National Center for Construction Education and Research (NCCER)**
http://www.nccer.org/carpentry

■ **The National Association of Home Builders offers Pre-Apprenticeship Certificate Training (PACT)**
http://www.hbi.org/Products-Services/Licensing/PACT